THE FARMING INDUSTRY

THE FARMING INDUSTRY

BY JILL MENKES KUSHNER

A GROLIER COMPANY
FRANKLIN WATTS
New York ▪ London ▪ Toronto ▪ Sydney
A FIRST BOOK ▪ 1984

Cover photograph courtesy of Shostal

Photographs courtesy of:
New York Public Library Picture Collection: pp. 5, 8;
Sunkist Growers Inc.: p. 15;
Delmarva Poultry Industry Inc.: p.18;
U.S. Department of Agriculture: pp. 21, 27, 34, 35, 42, 43, 49, 60, 64;
Culver Pictures: p. 31;
Supermarket News, Cities Media, 1984: p. 55;
Care: p. 71.

Library of Congress Cataloging in Publication Data

Kushner, Jill Menkes.
The farming industry.

(A First book)
Bibliography: p.
Includes index.
Summary: Traces the history of farming in America, describes the different kinds of farming, and discusses farming as a science and a business.
1. Agriculture—Economic aspects—United States—Juvenile literature. 2. Agriculture—United States—Juvenile literature. [1. Agriculture. 2. Farms. 3. Industry] I. Title.
HD1761.K87 1984 338.1'0973 84-7406
ISBN 0-531-04822-5

Printed in the United States of America
6 5 4 3 2 1

CONTENTS

THE FARMING INDUSTRY

1

THE HISTORY OF FARMING

Farming is the most important occupation in the world, simply because we need food in order to live. And, for a long time, people have tried to find more effective ways to gather, grow, and store food. In other words, they try to be better farmers. So, the history of farming goes back a very long time.

For our purposes, it will be easiest to look at farming by keeping certain stages in mind. These stages overlap in some cases, but they provide some sense of the enormous steps that people had to take to develop better approaches to farming.

The first stage occurred before the beginning of farming. Primitive people were concerned, first and foremost, with their own survival. They took the food that was available to them at the moment, but they did not *plan* or think about how they would get food for their future, as farmers do.

So, in the earliest days of mankind, people simply found their food on bushes, wild plants, and trees. They made some crude

STAGE 1

gathering food
hunting and fishing
making tools and weapons

STAGE 2

taming wild animals
preparing the soil
growing plants from seed

STAGE 3

preserving food
using animals for plowing
irrigating, rotating crops, and fertilizing

STAGE 4

inventing machines
learning, investigating
applying new methods

weapons and tools, which helped them to add fish and meat to their diet. But these activities focused on day-to-day survival, not on how to use the land to produce more food for the days, months, or years to come.

The first farmers were the people who learned to tame wild animals. They created fences to keep the animals from wandering and built homes for themselves so they could be near the animals. This led to a less nomadic way of life because people could raise several generations of animals in captivity to feed their families. This innovation marked the beginning of the second stage of the development of agriculture.

Along with the discovery that animals could be raised came the realization that plants grow from seeds. Early farmers probably found this out by observing seeds which were blown from trees or plants onto the ground. The seeds took root, sprouted, and became the plants or trees they had originally come from.

When farmers realized this, they began to plant wild grass seeds, which grew and provided food for their animals. In order to prepare the soil—to make it rich so that the seeds would grow well—farmers would cut and burn tree bark. The ashes of the bark would enrich the ground. This method is still used in some countries.

The earliest records we have show that farmers were thought to have lived over eight thousand years ago in the Middle East, in India, Asia, and in Mexico. In North and South America, Indians were developing their own farming methods.

Probably women first cultivated plants while men hunted and fished. Families began to live in clusters, helping each other with the daily work. They were able to learn from one another, to grow more food together than they could alone, and to share their food with one another.

As this process grew, people started improving their farming methods. New tools were created. Fresh food was preserved for later use by drying in the sun or by fire. Stone and clay pots were made to store food. Animal skins were used to make containers, clothing, and shelter. In "Stage 3," a still more settled life began.

The settlements of this period became more permanent; the first cities emerged. The two first important civilizations were in the Nile River Valley of Egypt and in Mesopotamia northeast of Egypt. Both areas were near rivers with good soil but not enough rain. These two regions were the first to be irrigated, in about 3000 B.C. By building canals that brought water to dry areas, it became possible to grow crops most of the year. Grains and milk were the earliest products.

In both Egypt and Mesopotamia, oxen were used to pull plows. Before that, plows had been pulled by human effort, and of course using an ox was much faster and easier.

During the next several thousand years, men and women from the Middle East moved to Europe and Asia. They cleared the new land and taught others their techniques. Wheat became the basic staple of man's diet.

By 200 A.D., the Romans had conquered much of Europe, parts of the Mideast, and the north of Africa. During the time of the Roman Empire many systems of improving the soil developed.

This scene from the wall of an Egyptian tomb shows how agricultural tools had developed by about 1,000 B.C In the lower portion of the picture, one man is plowing while others sow seeds and use scythes to harvest crops.

The Romans thought of the idea of leaving half a field fallow, or unplanted. This unused soil kept in needed nutrients and water for use in the next planting season.

Roman farmers used another important method of rotating crops. They planted peas and beans as one crop, alternating them with the main crop, because this would provide the soil with added nitrogen. They also built irrigation canals and large storage buildings.

Also during this time, the Romans performed the rather advanced feat of what we now call selective breeding. That is, they mated animals with especially good characteristics, and they saw that the new generation of animals displayed these same good qualities. In this way they produced the first special breed of dairy cow, now commonly known as a Holstein.

In about 400 A.D., the Roman Empire was invaded. When it fell, the thousand-year period known as the Middle Ages began. Farms became part of larger estates called manors. Lords who ruled the manors did not allow the peasants who worked on the farm to own land.

During the Middle Ages the system of crop rotation was improved. Instead of half the land being left unplanted, one-third of the land was left fallow, unplanted, but two-thirds could be farmed.

In the 900s a new harness was created to hitch horses to plows. Horses then began to replace oxen as the animal used for farm labor, which was a helpful development since they could work at least three times faster than the slow-moving oxen.

When the Middle Ages came to an end, peasant farmers received better wages, but it was not until the 1700s that they could own land in many parts of Europe.

European explorers of the 1400s and 1500s brought back from the New World much new farming lore. The Indians taught Europeans how to grow such crops as corn, peanuts, potatoes, and tobacco. From Europe the explorers brought seeds, livestock, and their own tools and methods, which they exchanged with the Indians.

In most of North America the early settlers established farms primarily to survive. But the 1700s saw the development of a kind of "agricultural revolution," brought about by improvements in crops, livestock production, and the invention of new farm equipment. Farmers grew so much food that they could feed their families and sell the extra food for profit.

The rash of new machinery that was developed in the 1700s and 1800s marks the beginning of our "Stage 4." An entirely new era was introduced, with high levels of production, and a new kind of speed and efficiency. Of the many inventions produced, the following were highly significant:

1701 ▪ Jethro Tull invented a *seed drill*, which allowed farmers to drill rows of holes in the soil for seeds. Before this time seeds had been hand-scattered. First machine with moving parts on the inside.

1793 ▪ Eli Whitney invented the *cotton gin*, which separated the cotton fiber from the seed. This made large-scale production possible and helped cotton replace tobacco as the South's chief crop.

1797 ▪ Charles Newbold patented a cast-iron plow.

1834 ▪ Cyrus McCormick patented the first *reaper*, a harvesting machine that cut grain quickly.

1834 ▪ The Pitts brothers patented a *thresher* that separated the grain from its husk.

1836 ▪ Hiram Moore and J. Haskell patented a *combine*, which cut and threshed at one time.

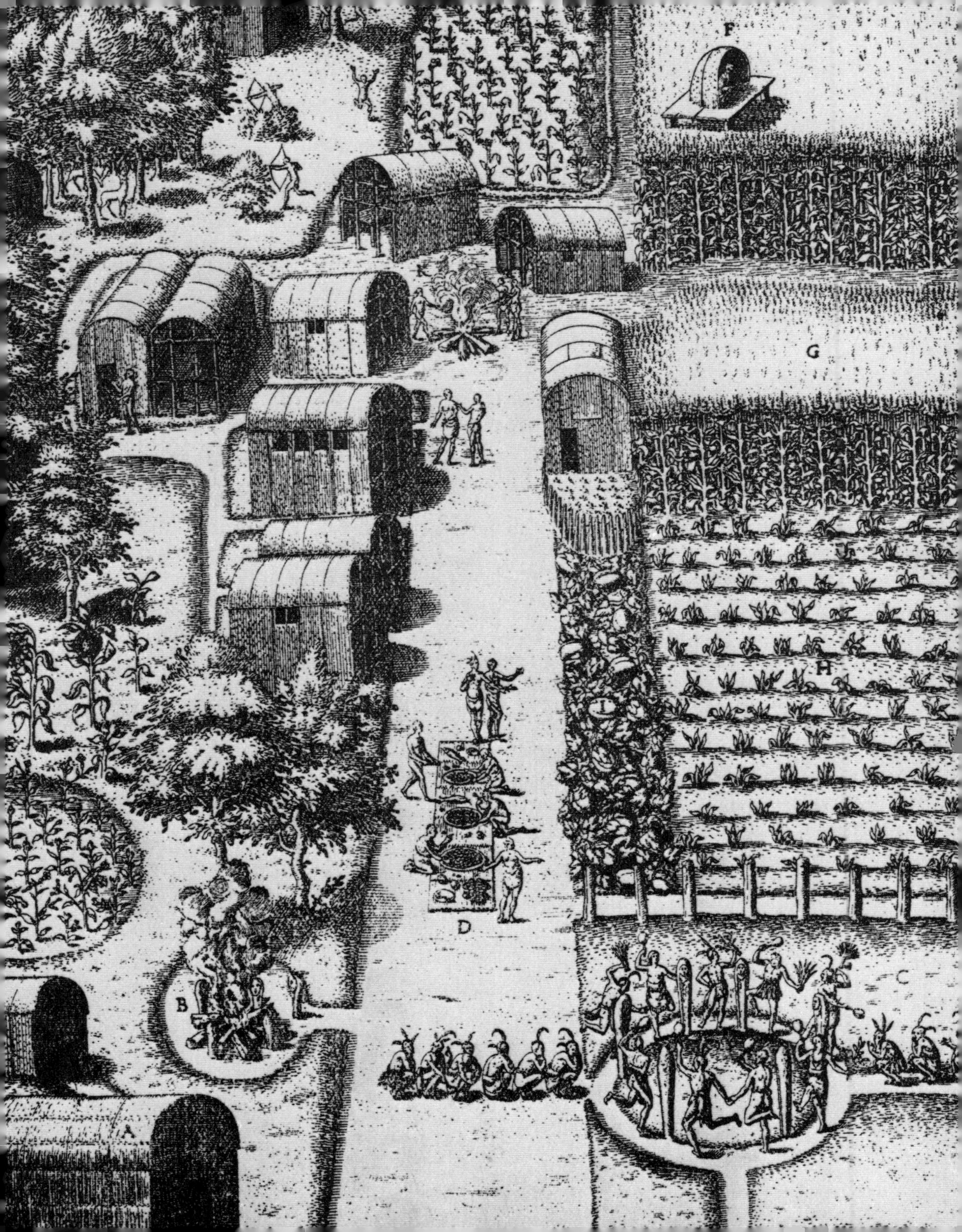
F
E
K
G
H
I
D
B
C
A

Farming in early America was affected tremendously by the new government's attempts to cope with such problems as land rights and ownership. Certain laws had a particularly large impact. Among these laws was the Homestead Act, of 1862, which granted land to settlers at a low cost, or at no cost if they lived on the land and farmed it for five years. The Homestead Act encouraged settling new land, especially in the West and Midwest.

Also in 1862, the Morrill Act was passed, granting large tracts of land to the states for the purpose of setting up colleges of agriculture, called land-grant colleges.

Congress also supported the building of railroads to improve food transport. The first railroad from the eastern states to the Great Plains was finished in the late 1860s.

As farmers learned new methods and improved their old ones, even newer approaches were discovered. In the Midwest a dry farming system was developed to help soil store moisture for the coming year. Wheat became an important crop, as farmers found it could grow well in dry climates.

Agriculture in the 1900s brought an emphasis on research and technology. The 1920s saw the first all-purpose gasoline tractors, able to provide power to a variety of farm machines.

Early explorers in the New World found Indians growing a variety of plants. Indians in this village, in what became the colony of Virginia, cultivated fields of corn (to the right of the road). A patch of squashes and melons is growing next to the corn, along the road, and on the other side of the road is a round field of tobacco and a few sunflowers.

The early 1900s heralded the development of new varieties of crops. Hybrid corn increased the yield of corn crops. This work was based on the ideas of Gregor Mendel who, in the 1800s, had done substantial research in the new science of genetics.

Electricity was not in wide use on American farms until 1935, when the federal government established the Rural Electrification Administration (REA). REA granted low-cost loans to farmers so that they could develop power. Electricity has proved to be a great boon to the farmer, who today uses it to operate everything from grain elevators to water pumps.

Better types of livestock were produced in the twentieth century; better nutritional and medical care were emphasized. Especially with dairy cows and poultry, improvements in breeding have greatly increased production.

Since the late 1700s, many chemicals had been used in farming. By the 1800s, certain elements, especially nitrogen, had been found to be important to the growth of crops. The creation of low-priced nitrogen in the twentieth century has meant a large increase in the amount of chemical fertilizer used.

Insecticides have long been used to control crop destruction. And herbicides, particularly the widely used "2,4-D", have been a great help in preventing weeds from choking growing plants.

Better control has been attained over plant and animal disease. The discovery of the causes of plant disease; the creation of new strains of plants that can resist disease; the introduction of animal vaccines—all of these have helped.

Up to the 1920s farm production increased chiefly because new farms and grazing lands were carved out of forest land. Because soil and water were thought to be almost limitless, they were not conserved. Between 1920 and 1940, however, there was a greater emphasis on improving and conserving the land.

The number of farms and the amount of land used for them stayed fairly stable, and increases in the level of production were brought about as a result of mechanization.

Since 1940, the number of farms in America has decreased while the size of individual farms has increased. Farmers now make up just 3 percent of our population, but they produce a tremendous amount of food.

In countries such as America, where farming has become so sophisticated, production has become so great that we require fewer farmers to feed our own people. We are able to feed people in other countries as well.

Some older farmers, who have watched the old ways disappear and new methods take their place, may long for a quieter, less complicated life. But the reality is that farming has become big business. Today, successful farmers are experts in finance as well as crop science; nevertheless successful farmers still possess the skill and courage to tackle a time-honored, if increasingly complex, task.

2

KINDS OF FARMING

In colonial times, when there were no machines and farming was done on a rather small scale, most farming was *subsistence* farming. Almost everyone, including government leaders, farmed for survival, or subsistence. Each farm family had some basic crops, such as grains or vegetables, and at least a few cows and chickens to satisfy their need for milk, meat and eggs.

Those settlers who were able to do so produced crops for sale, in addition to what they used for their own needs. These *cash crops* were sold locally and for export to England, Spain, and France.

In modern America, the cash-crop and livestock-for-cash farm prevails. Today, farming is much more a business than a means of feeding one's family. The United States has an extremely successful farming industry. In fact, our agricultural assets—our natural resources and up-to-date scientific and business knowledge—have contributed to America's importance in the world.

Subsistence farming deserves a brief mention, because it is part of our history, but primarily because many nations in the world still practice it. Farmers in Africa, Latin America, and Southeast Asia still farm primarily to meet their own food needs, not to sell to others. In these areas, which have large populations and little farm machinery, farms are worked chiefly by human labor.

Although we in the United States are fairly far removed from subsistence farming, it is important for us to understand that large groups of people in the world still work as subsistence farmers. As we try to share our knowledge of modern agricultural techniques with others, we need to be aware that we are introducing methods that are completely unfamiliar to them.

In our country, about half the land, or 1 billion acres, is used for crop or livestock production. Of that amount, 450 million acres is cropland, more than any other country except the Soviet Union. We grow 64 percent of the world's soybeans, about 50 percent of its grapefruit and corn, 25 percent of the world's beef, and 20 percent of the world's grain and eggs. These facts are especially impressive when you consider that we have fewer than 0.3 percent of the world's farmers.

Over the years, certain areas of the United States have become known as specialists in particular products: the Midwest has been known for corn and hog production; the Great Plains and Northwest regions, for wheat; the South and Southwest, for cotton; and an area from New York to Minnesota, for dairy products. Although certain items are heavily produced in given regions, improvements in irrigation and livestock and plant breeding now make it possible to produce many crops in additional regions.

Several factors influence a farmer trying to decide what type of farming to attempt. Physical factors, such as the climate, topography, and the soil structure, are all important. *Climate* refers to the length of the growing season, the amount of rainfall, and how moderate the temperature is. *Topography* is the kind of land: mountainous, level, or rolling hills. *Soil structure* must be determined to find if the soil is workable.

Other considerations have to do with the biology of the region. The kinds of insects, weeds, and diseases that thrive in an area are all issues to think about. Even though farmers can deal with many problems by using modern chemicals or equipment, they still have to decide whether it is worth the cost of controlling whatever may damage the product.

The economic decisions a farmer must make are many. For instance, in certain parts of America, huge numbers of farm laborers are needed to help pick fruits and vegetables by hand. So the availability of inexpensive labor is important. Other things to consider are the location of nearby markets for perishable foods such as fruit, vegetables, milk, or eggs.

WHAT FARMS LOOK LIKE

Farms look almost like communities in themselves, with farmhouses and several other buildings as well. Some of those buildings are used to store machines such as tractors, plows, balers, and pickers.

Here's a brief description of certain types of farms. You would certainly notice other things as well if you were to visit one of them.

Lemons, like all citrus fruits, require a warm climate for successful cultivation.

Dairy farms have:

- a large fenced-in pasture for grazing
- an open barn with milking parlors
- at least one silo to store grain for the cows' feed
- a cooling tank
- fields of corn or alfalfa for feed

Grain farms have:

- fields of crops
- feedlots where livestock are fed (if it's a mixed farm)
- storage and drying areas for grain
- large buildings to store harvesting machinery

Fruit farms have:

- rows of trees
- boxes and baskets for packing fruit
- refrigerated warehouses
- smudge pots kept available in case of frost
- spraying machines to kill insects

Poultry and egg farms have:

- a chicken house with narrow windows, filled with cages, some with sloped floors so that eggs can roll down the floor onto a conveyor belt
- inspection room, receives eggs from chicken house
- outside living areas if there are turkeys or ducks

The farms that we have mentioned are all called *specialized* farms because they produce one or two main crops or kinds of livestock. Specialized farms use mass-production techniques and equipment. For example, a farm specializing in egg production may have a conveyor belt to move the eggs from the chicken house to the inspection room. Most American farms—95 percent—are specialized. About half of these farms produce crops, and slightly fewer than half produce livestock.

■ Specialized crop farms are also called one-crop or cash-crop farms. Cash-crop farms are those that produce wheat, as in the Great Plains; or plantations, largely found in the South, that grow mainly one crop such as rice or sugar cane.

■ Specialized livestock farms are feedlots, poultry and egg farms, and dairy farms. In the far West, huge ranches are used to raise cattle and sheep. When the cattle are between 5 and 12 months old they are sent to the feedlot. There they are kept in pens, where they are fed grains and high-protein feed that will make them gain weight before slaughter.

Other examples of livestock farms include special farms for horses, mules, and even bees and fish. All of these are called livestock "farms," even though they don't all look like farms.

■ Diversified, or *mixed,* farms are those that produce several crops and livestock. They comprise just 5 percent of today's farms, although once they were the main way of farming. With a diversified farm, a farmer could be sure of one crop if another failed. Also, raising both crops and livestock enabled him to feed his own animals instead of having to buy feed.

As we mentioned, most of our farms are now specialized, due to many factors. There are fewer risks of crop or animal loss because of the discovery of new chemicals that fight disease and

pests. Also, if the price of a given crop drops in the marketplace, there are not as many risks as there once were. Today the government helps farmers stay economically stable.

Perhaps the largest contribution to the rise of specialized farms has been the demand of the marketplace. Americans tend to like bulk production: that is, a constant supply of high-quality, low-cost products. So the farmer who does one thing and does it well has an attractive product to offer for sale in the marketplace.

As many as 20,000 chickens grow to the market size of about four pounds in this ''broiler house'' on a poultry farm.

3

CROPS

American farmers grow a wide selection of crops. This variety is a result of our great range in climate. However, every crop is produced according to certain guidelines.

Crops must be planted in the proper soil, and supplied with the ingredients that provide a nourishing environment. The soil also stores water, which is drawn up into the plant through its roots. In the early twentieth century, scientists discovered that ten elements were essential in crop soil: carbon, hydrogen, oxygen, nitrogen, potassium, phosphorus, sulfur, magnesium, iron, and calcium.

The most fertile soil is the surface or topsoil. Topsoil is light and has to be protected from being washed or blown away.

The farmer first must decide what sort of *tillage* method to use. Tillage involves breaking up the soil to make it ready for planting. It is needed because the soil structure changes with every planting and harvesting; it needs to be brought back to its original state.

A farmer is plowing a strip of land where he will plant corn. To protect his land against overuse he will leave alternate strips of grass between strips of corn.

- Primary tillage uses several different kinds of plows to break up soil:

    - a *moldboard plow* has a sliding bottom and is used for many different kinds of soil.
    - a *disk plow* uses round steel disks. It is good for deep plowing and works well in both dry and moist soil.
    - a *rotary plow* has a set of knives that rotate on a shaft. It requires extra power equipment.
    - a *chisel plow* is used to loosen hard, dry soils.
    - a *subsoil plow* digs under the surface, especially when better drainage is needed.

- Secondary tillage improves the planting area by breaking up the soil further. The equipment used—harrows, rollers, and pulverizers—does not penetrate the soil deeply.

- Other tillage methods such as *reduced tillage* and *no-tillage* sometimes use mulches spread over the topsoil. These methods have gained some popularity because they help conserve soil and water. Too much tilling ruins the soil and makes it less able to store water. There are also special plows that conserve soil by not turning it completely over.

After the soil is ready, it is time to plant. Planting is usually done in the springtime, when there is no danger of frost.

Machines called *drills* cut rows (furrows) in the soil. Modern drills also drop seeds into the furrows, cover them with soil, and even apply insecticides and fertilizer. Many rows can be planted at one time.

Farmers rely on either rainfall or irrigation (See chapter 5) for water. If rainfall is light, farmers can also use dry farming, which means that available moisture is used without irrigation.

After planting, weeds must be eliminated. *Harrows*, which are plows with sharp metal teeth, uproot weeds that have not been destroyed in the original plowing, or those that have come up after weed-killers have been applied.

When the crop is ready for harvesting, many machines are used. Modern *combines* are used to harvest grains such as wheat, oats, barley, and corn. Other specialized harvesting machines are cotton-pickers and potato-diggers.

Field choppers and silage cutters cut plants for feed. These machines collect and chop up plants, which can then be used as feed, or stored in a silo to be used for feed at a later time.

Once harvested, foods such as wheat, corn, and rice require special drying equipment and huge storage bins. With other crops, such as fruit, refrigeration and careful packing are important. Forage is stored dry (hay) or kept moist (silage), according to the animal it will be used for. The food is stored until it can be sent to the various markets for sale.

MAJOR FOOD CROPS

The crops of the world are a vital food source for human beings. They provide food for us to eat, and they also feed our other main food source, the animals we eat.

Corn ▪ Corn has been part of American agriculture for many years. The Indians of North and South America used it to make corn meal and flour; they developed between 200 and 300 strains of corn.

Corn has always been a strong crop, resistant to weather

and disease. It has been cross-bred many times to produce several varieties with more nutritional qualities. It is used nowadays for cereals, cornmeal, snack foods, oils and syrups, as well as for livestock feed.

Wheat ▪ Americans eat more wheat than any other food. There are more than two hundred kinds of wheat produced in the United States.

These varieties are classified by texture, color, and style of growth. For example, "hard" wheats are used to make bread. "White" wheats are used for spaghetti, macaroni, and other pasta. "Soft" wheat is used for baking flour for cakes and cookies.

Soybeans ▪ Soybeans were first used in America as a rotation crop. They enriched the soil by providing nitrogen to be used by other crops.

Production of soybeans has increased greatly over the last fifty years. During the past twenty years, scientists have managed to improve their texture so that soybeans have become a more appealing meat substitute.

Soybeans are also used to make margarine, mayonnaise, salad dressing, cooking oil, soy sauce, and soy cheese (tofu). Soy milk is used for infants and for others who are allergic to cow's milk.

Soybeans are an especially flexible crop. They can be planted later in the season than crops such as corn and survive with less rainfall.

Oats, Barley, Rye ▪ In the United States, grains such as oats, barley, and rye are grown for special purposes. Although high in protein, oats are not used as commonly as wheat in food for humans. In America both oats and barley are used mainly for livestock feed. Barley is also used in soups and for the malt in beer.

Although rye is not generally an important crop, its production is increasing because rye bread is becoming more popular. As people become more aware of good nutrition, they are beginning to eat breads such as rye, which contain good natural sources of vitamins.

Potatoes ▪ Potatoes, which are inexpensive and easy to prepare, are a staple food for many people. Always an important crop, they are undergoing further expansion in the modern market due to the introduction of more processed foods. Today consumers can buy frozen french fries, potato chips, and dehydrated potatoes.

Rice ▪ Rice is the world's single most important crop. About half the world's people depend on rice; Asians consume it as three-quarters of their diet. Many Far Eastern cultures hold rice in sacred honor. In Japan, it is valued as being second in importance only to the Emperor, and to waste rice is considered a sin.

Rice, which is one of the world's oldest foods, was brought to the New World by Columbus. By the late 1600s this tough, high-yield plant was established as a thriving crop in the southern colonies.

Today in the United States, most rice is processed. Our processing methods are very efficient in comparison to other countries such as India and China, where rice is usually hand-cultivated just as it was thousands of years ago.

NONFOOD CROPS

Nonfood crops such as tobacco, cotton, and flaxseed are important for providing needs beyond food, such as clothing, shelter, and even simple enjoyment. For cash-crop farmers, these items have represented large profits over the years.

Tobacco has been widespread in both Americas. The Indians have a long-standing tradition of smoking tobacco at official councils, called "passing the peace pipe."

Tobacco was grown for profit as early as the beginning of the 1600s. It was America's main export until the 1800s, when "King Cotton" took over as the most profitable crop. Today, its popularity has declined as people are more aware of tobacco-related health problems.

Cotton, one of the oldest existing fibers, has been used for clothing and shelter for thousands of years. American cotton production was revolutionized by Eli Whitney's invention of the cotton gin ("gin" for "engine") in 1793, which made harvesting go fifty times faster.

In the 1930s, boll weevils destroyed many cotton crops, and synthetics such as rayon became popular. Today cotton is once again a popular item. Picking is now done completely by machine; insecticides have controlled most pests; and new cotton fabrics have been developed which, like synthetics, are wrinkle-free. Many parts of cottonseed are used to make margarine, livestock feed, fertilizer, and even plastics.

Flaxseed, also called linseed, is used to make linseed oil, varnishes, soaps, and linoleum. Brought to the colonies in the 1600s, flax was originally used to make linen, but its popularity in the United States declined as the use of cotton became more widespread.

Cotton harvesters at work in a Mississippi field. Each machine can pick as much in one hour as a person could in a week.

FRUITS AND VEGETABLES

Fruits and vegetables are produced most often in wealthier countries, because they are less of a staple than cereal or meat. In America we have an abundance of both fruits and vegetables. Fruits are generally grown in orchards on large farms. Vegetables are not only grown on large farms, but also on small truck farms located near marketplaces for ready sale.

Citrus fruits—particularly oranges, grapefruit, and lemons—are the most popular fruits in the United States. They grow mostly in California and Florida. There are over twenty-five different kinds of oranges. Valencia oranges, for example, are used for juice, navel oranges are eaten plain, and other varieties are used for making marmalades and other foods.

Although apples are not the favorite fruit in the United States, they are the most common fruit in the world. In America, a man named John Chapman planted apple seeds all over the midwestern United States and became known as the legendary Johnny Appleseed. Apples are eaten raw and used in cooking, and to make apple cider; the best-known kind in the United States is the Red Delicious.

Although vegetables are not as economically important as other American food crops, they are still essential products. California, Illinois, Iowa, and Minnesota are the largest vegetable-producing states in the United States.

Various parts of a vegetable become foods that we consume. For example, there are bulb vegetables (onions), flower clusters (cauliflower), fruits (cucumbers, snap beans), leaves (lettuce), roots (carrots), seeds (peas and corn), stems (asparagus), and tubers (potatoes). Tomatoes and lettuce are the most abundantly produced vegetables.

Tomatoes, which are technically a fruit, were once used

mainly in sauces; they were thought to be dangerous to eat, even poisonous. Today, of course, we enjoy tomatoes in all forms, and many people grow their own in home gardens. A variety of sizes are used for eating, ketchup, pastes, sauces, and juice.

Lettuce is a very popular vegetable, grown in three main forms: head lettuce, loose-leaf, and butterhead. It is used mainly in salads. Lettuce has well-known relatives such as cabbage, broccoli, cauliflower, and parsley.

4

LIVESTOCK

The taming of animals dates back thousands of years to the Middle East, Greece, Asia, and parts of Europe. Sheep, goats, and cattle were the earliest farm animals.

Much early American livestock was imported from the Old World and the building of a real industry was slow and difficult. In the 1800s, livestock was raised in the West, then driven overland to markets in the East. In this way, the animals acted as their own transportation, but they lost weight during the trip and their meat was not very tender.

Farmers became more aware of the needs of their animals in the mid-1800s. They began to shelter the animals against the cold and gave them fodder throughout the winter. Farmers also provided more nutritious feed, a quieter environment, plenty of water, and more careful handling.

Important advances in the livestock industry include refrigeration, which allowed meat to be stored for long periods of time.

Cattle in the Kansas City stockyards around 1895

The first refrigerated transport in 1865 took the form of rolling ice-boxes that contained artificial ice.

Other major developments of the late 1800s were Gregor Mendel's work in genetics, which led to advances in animal breeding; the discovery of organisms that caused disease; and treatments for major livestock illnesses such as malaria, yellow fever, tuberculosis, and tick fever.

In addition to these important steps, the development of immunizations against disease, as well as greater emphasis on cleanliness, resulted in healthier livestock.

Since 1900 the number of livestock products has more than doubled. There is much more emphasis placed on selective breeding—intentionally breeding stock to produce superior offspring. For instance, agricultural experts have experimented in breeding to produce calves with improved characteristics such as high milk production. Also, the U.S. Department of Agriculture now has disease-control centers in port cities. Recently, the USDA was able to stop the spread of foot-and-mouth disease by keeping imported cattle in such a center.

Livestock production is now quite sophisticated. Many modern calves are raised on ranches, brought to feedlots to be fattened, and shipped to market when they weigh about 1100 pounds. Feedlots often use hot meals and a computer determines necessary feed ingredients.

Marketing of meat, once located primarily in Chicago, Omaha, and Kansas City, now takes place nearer to producing farms. For example, in Texas many farms not only produce cattle but are also involved in processing meat.

CARE OF LIVESTOCK

The basic care of livestock means providing food and shelter, and making sure the animals are healthy. It sounds pretty simple, but it involves a lot of hard work for the farmer. For example, all the following factors must be considered.

Food ▪ Cattle and sheep eat quite a bit of *forage*, which are grazing plants, hay, or silage. Forage is good for their digestion. Both animals are pretty efficient in that their bodies can convert the forage easily into the protein they need.

Livestock also eat *feed concentrates*, which contain grains (corn, sorghum, soy meal) mixed with vitamins and minerals. Feed concentrates supply energy but not much roughage. Hogs and poultry are given only concentrates, since they are unable to digest forage well.

Over half of America's farmland is used for grazing, mostly for cattle and sheep. Sheep being raised for their wool live on grass and stay on the ranch. Cattle graze until they go to feedlots. Dairy cows graze too, but farmers use feed to help increase milk production.

Livestock are fattened both by grazing and with feed concentrates near where grain is produced. Some farmers fatten livestock on their own farms, but many send the animals out to feedlots for this purpose.

Housing and Confinement ▪ Mature beef cattle and sheep are kept outside all year, as they are not terribly affected by the cold. Younger livestock need some kind of protection such as sheds, and some are raised entirely indoors (for example, poultry).

Poultry and egg farms are good examples of confinement. On a confinement farm, animals are kept in small areas, moving around hardly at all. Using less energy, chickens become fatter and hens produce more eggs.

Health and Breeding ▪ Medical care of animals has become much easier in the twentieth century. Vaccines have played an

*Sheep graze on
a California ranch.*

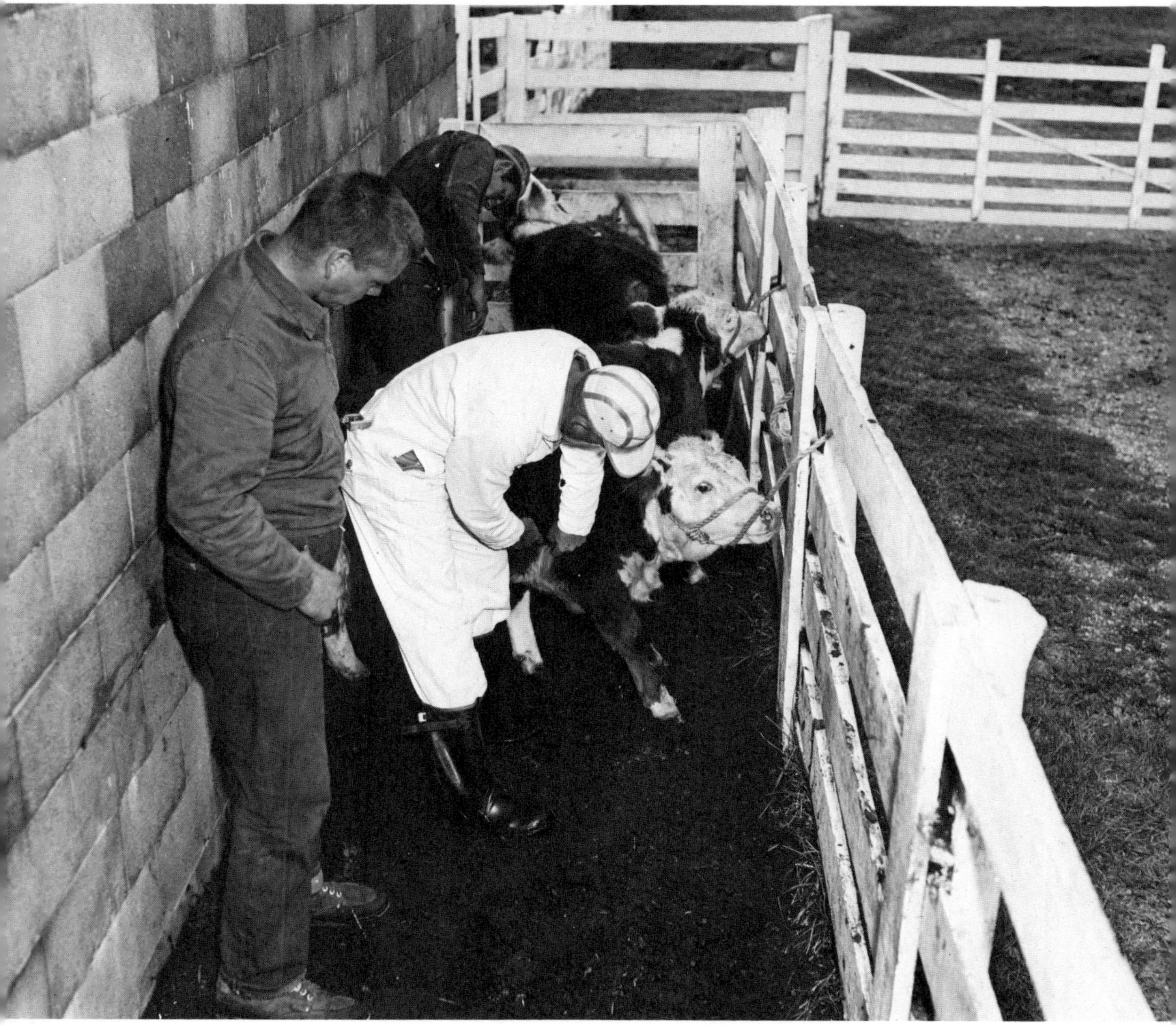

A veterinarian vaccinates a calf on a cattle ranch.

important role in eliminating disease, and the development of medicines such as penicillin has helped to treat sick animals.

Most livestock is raised for meat, but some animals are kept as *breeding stock*. These might include a cow who produces excellent milk, or a hen who lays many eggs. Breeding stock are used to produce offspring who will hopefully also be superior. When animals are used in this way, the process is called *selective breeding*.

Although cattle are by far the most popular meat source in America, the development of hogs as valuable livestock is on the rise. They require little shelter and are very inexpensive to raise, since they are able to eat the seeds of many different plants.

Lamb, the meat of young sheep, and mutton, the meat of older sheep, are not as popular in the United States as in other countries. Their most important product in this country is wool.

DAIRY FARMING

Dairy farms flourish in cool climates with good rainfall. They are usually located near the cities where their products are sold and consumed. In colonial times, dairy products had to be used soon after they were made, since there was no refrigeration and transportation was so slow.

In the late 1800s and early 1900s, several changes helped to modernize dairy farming. In the 1860s, Louis Pasteur removed the bacteria from milk by boiling it. In 1873, the first silo was built, allowing farmers to store grain which they could feed to their cows all year round. Transport was improved with the first refrigerated rail cars in 1875, and by the early 1900s, milking machines were being used. These new developments allowed products to be produced in greater quantities and sold in faraway places.

Another important discovery was the realization that dairy cows are sensitive animals. They did not do well in the cramped milking stalls that were used for many years. They suffer from the cold and require a lot of personal attention. Dairy farms are therefore kept small, and cows are exercised outside, and brought in to be milked in individual stalls.

Another concern for dairy farmers has been the cost of feed. Feed comprises almost half the cost of an operating dairy farm. Thus, dairy farms are very much affected by the market price of grain.

Other important dairy products are cheese, butter, and yogurt. Butter, however, has been in less demand in recent years because of its high cost and because it contains cholesterol, which doctors have found contributes to heart disease. In the United States, cheese is widely consumed as a meat substitute and as a snack food.

POULTRY FARMING

The first commercial incubator was used in 1885. Since the incubator provided an evenly maintained temperature, eggs could be kept warm enough to hatch, and more newborn chicks survived.

Today most chickens are kept in small areas where they are fattened with a special diet. The assembly-line approach has worked well in chicken farming, leading to a decrease in the time needed to raise a broiler chicken, and therefore to an overall lessening of costs.

In the 1950s, most Americans bought whole chickens at the local butcher shop. Today we buy chickens from a supermarket, with our favorite parts already cut and packaged.

People also enjoy eating eggs, and today we buy them in supermarkets as well as fresh from the farm. Recent egg consumption dropped because of concern about cholesterol content, but eggs are still considered a useful protein source.

5

THE SCIENCE OF FARMING

Farming has always required special knowledge, combined with good intuition. Farmers need to have a good "sense" about the work that they do, but scientific advances have helped to make farming into the tremendously productive enterprise it is today.

We will deal with three broad areas that have influenced farming: *mechanization*, the invention and improvements of machines and energy sources; *biology and chemistry*, plants, animals, and the use of chemicals in farming; and *research*, the information that has been collected to aid farmers.

MECHANIZATION

Only a few hundred years ago, farmers worked with sharp sticks, hoes, rakes, and spades. When a farmer used a work animal such as an ox with his plow, he could plow one acre in a day. (An acre is 43,560 square feet, or one-third the size of a square city block.)

Today it takes less than one hour to plow an acre. Tractors and other machines have completely revolutionized farming.

Whereas in 1870, over half the U.S. working population was employed on farms, in 1982 fewer than three out of every hundred Americans were farm workers. Yet they produce much more than ever before. This is due largely to the advent of machinery, as well as the improvement in kinds of livestock and crops.

The basic farm tool until the American Revolution was the plow, made of wood and quite crude. The first improvement of this tool was to add iron and, later, steel parts for the cutting blades (shares) on the plow bottom. With the development of the tractor, plows were able to move far more quickly.

The first man to operate a gasoline tractor successfully was John Froelich, in 1892. In 1905, C. W. Hart and C. W. Parr formed the first tractor-making business. Early in the 1920s tractors began to replace horses, and by the 1960s there were five million tractors in use.

Other machines have been developed for all sorts of farm purposes. Specialized machinery reshapes the land to make it more workable. A variety of plows, depending on the nature of the land, are attached to tractors to till the soil. Then, planting machines place seeds in several rows at once.

When the crop is ready, harvesting machines do their job. Harvesters with tree shields protect fruit trees from injury during picking. Corn-pickers remove whole ears from stalks and specially equipped combines roll through the field separating kernels from the cob.

Haying has been modernized so that the hay is now raked, cut, piled, and crushed automatically. Cotton is also mechanically collected; strippers and pickers separate the cotton boll from the plant, and the seeds from the fibers.

In addition to land vehicles, airplanes have been in use on

farms since the 1920s for dusting crops with insecticide. Today they are also used for fertilizing, weed-killing, and planting.

An economic look at the past twenty years tells us part of the story of farm machinery: In 1960, farmers spent over $1 billion on mechanical farm equipment. In 1980, farmers spent over $7 billion. Even accounting for inflation, there is no questioning the enormous role machinery plays, and will continue to play, in the work of the farmer.

Another important addition to farm life has been electricity. As recently as the 1930s, only a small percentage of American farms were electrically powered. With government assistance, farmers were able to get low-interest loans to power their homes and other farm buildings.

Electricity has increased the quantity and quality of farm output. Dairy farms have made great use of it for milking machines, refrigerators to cool and store milk, water pumps to clean cattle stalls, and electric transporters to fill feed bins. One farmer can now milk as many as 120 cows within two hours.

Poultry farmers have benefited too. Egg-hatching, warming of baby chicks, and sorting and grading of eggs—all of these tasks are now performed with the aid of electricity.

BIOLOGY AND CHEMISTRY

Knowledge about the biological make-up of plants and animals, as well as about chemical usage, has proved beneficial to farmers. The interaction between biology and chemistry is very important in agricultural science. Scientists called biochemists study such subjects as erosion, cropping methods, chemicals, irrigation, breeding, feeding, and disease control.

We now realize that early farmers were unaware of the toll

that farming takes on the land through erosion. There are two types of *erosion*, or wearing away of the soil. *Wind erosion* occurs when the soil is so dry that it simply blows away. Plant roots will not hold. *Water erosion* occurs when rain washes off the topsoil. Many necessary nutrients are then lost.

Different kinds of planting methods are useful in preventing erosion. Contour farming allows for planting horizontally along the curves of slopes. Strip cropping, or planting rows of clover or grass between crops, can also help by slowing water run-off.

Crop rotation is an ancient method of conserving farmland. It helps to maintain the soil structure and reduce erosion. A farmer alternates clover or peas with his regular crop. These two plants do not remove nutrients from the soil, and they can replace some that are needed.

Another way to add nutrients to soil is by using fertilizer. Many chemicals have been used in farming for years, costing farmers a total of over $9 billion in 1978. Fertilizer in particular is very widely used.

Soil needs are determined by laboratory testing of soil and plant tissues. Then fertilizer is applied: spread over the ground as a liquid, inserted as a gas, or on a tape laid out through the crops, as seeds are planted. There are some problems with the use of fertilizer, as it does change the natural make-up of the soil. Researchers are now recommending the use of organic material in combination with the chemically produced fertilizer to protect soil.

Chemicals are also used to control insects and weeds. Once weeds were eliminated by being dug out of the ground or crowded out by crops. In the 1940s a compound known as "2,4-D" was invented. This and other herbicides, or weed-killers, are now used with little concern, since each chemical seems to kill just weeds without damaging cultivated plants.

The dramatic curves in this photograph are an example of contour farming.

Today, airplanes are used to spray insecticides and herbicides, which help increase the farmer's production.

Pesticides, or chemicals to control insects, have presented a problem. During World War II, a Swiss scientist developed DDT, which was used widely for insect control. Unfortunately, it did not break down easily, and thus meant a danger to human and animal life. Its use was restricted by the federal government in the 1970s.

The use of insecticides poses a pollution problem because the chemical is carried from one place to another by air and water. Farmers still use insecticides, but they also use other methods of insect control. One such method is to use insects' natural enemies, such as ladybugs or praying mantises, to destroy the pests.

Irrigation, long in use, has enabled farmers to bring water to dry land. Thus, entire areas of dry farmland have become more usable. There are many types of irrigation. The entire crop can be flooded; individual rows can be watered; and crops such as fruit trees can even receive water one "drip" at a time.

Since rainfall is not always predictable, the management of water is important. Dams and reservoirs help capture and hold water. Also, treating polluted water to make it clean makes it available for continued use.

Some great strides have been made in the breeding of plants and animals. Corn is a good example of *crossbreeding*—combining two varieties of a species—since almost all corn that we eat is a hybrid, or mixture. These new combinations have improved qualities such as better taste or better resistance to disease.

Some of our best cattle result from selective breeding, discussed in Chapter Four. *Inbreeding* is the mating of animals of the same kind or breed; *outbreeding* is crossbreeding from different kinds of cattle. Constant research is being done in this area, and also in *artificial insemination*. In this procedure, the animals do not mate. Rather, sperm is collected from the male and injected into the female.

In addition, knowledge about animal nutrition has increased greatly. Since 1930, many advances have been made which take into account the functions animals perform and the differences in their body structures. The digestibility and protein level of feed are also very important. Farmers pay special attention to animal nutrition, as it greatly affects their profits.

Livestock are victims of both internal and external pests, such as worms, flies, ticks, and lice. These can all be controlled with medicines; other insects are controlled by the spraying of crops which animals eat.

When disease erupts, however, it can be quite costly in time and money for a farmer. He must stop the disease before it spreads to the farm's entire livestock population. Veterinarians, who can diagnose and treat diseases, are extremely important to livestock farmers.

RESEARCH

The discovery of machines and of new biological and chemical approaches to farming would not have happened without the long and difficult work of research. Research has always been important to farmers. By the early 1900s, agricultural colleges and county extension services were set up to emphasize the importance of constant learning. Today the U.S. Department of Agriculture conducts studies and publishes information to help farmers.

Modern research is going on in many areas: establishing controlled environments for crops and animals; modifying the soil, water, and air; using climate-controlled greenhouses; creating plastic mulches; hydroponics (soil-less planting)—these are just some of the improvements being explored.

Research is also being done on problems of immediate concern to farmers. Among these are the way soil holds water differently during hot weather; the effect of outside temperature on dairy cows and beef cattle; the effects of weather on the application of pesticides and herbicides; and how to use forecasts to determine the best time for applying chemicals.

The ability of a farmer to observe and pursue new methods greatly affects levels of production. A farmer has to be always aware and learning. Moreover, the farmer must often weigh the risk of a large financial investment against the possible benefits to be gained by exploring new methods. He must have a high degree of skill, as well as good common sense, to perform his job successfully.

6

FARMING AS A BUSINESS

As we have noted, farming was once mainly a style of living in America. Today it is a complicated and challenging business. Consider these facts:

- Agriculture is our nation's biggest industry. It represents 70 percent of all of the profits earned in manufacturing.
- The agricultural industry is America's largest employer, with 23 million workers. Of these, 3.4 million work on a farm. (Others are involved in other parts of the farming process, such as working in a canning factory or a meat packing plant.)
- One farmworker now supplies enough food for 78 people.
- The United States exports more farm products than

any other country. These exports help our nation's balance of trade, because the profit we receive from the sale of these products helps to offset the money our country spends abroad on other items.

- Consumers spent $285 billion for farm products in 1982. But $200 billion of that went for processing procedures such as canning, freezing, packaging, and storage.

Over the last thirty years, there has been a rapid decline in the total number of farms in the United States. The number of large farms is growing, and the number of small farms has decreased. Some people worry that corporations, or big businesses, will start running farms, and that the days of the family farm are over. Actually, most farms are owned by families in the form of individuals, couples, or partners. In this chapter we take a look at those farmowners and at their business concerns.

In most cases, the people who own a farm also work on it. Farmers hire workers, part- or full-time, to help them, especially at harvest time to bring in the crops. But if you were to visit a farm, you would find that most of the farmer's family is also busy working.

The farmer must decide first what particular crops or livestock are right for the farm, and how much to produce. Soil, climate, and the cost of transporting goods all have to be consid-

This grain elevator is owned by a group of farmers called a cooperative. Members of the association use the elevator to store their wheat before it is sold.

FARMERS
CO OP

ered. If prices change, the farmer must be flexible enough to be able to change to a lower-cost crop or livestock product or one that is more in demand.

COSTS AND INCOME

Costs (expenses) are the amount of money a farmer needs to spend. Then he receives payment when his product is sold. To figure how much money he actually makes as *income*, he subtracts his costs:

Gross (total) income	—	$10.00 worth of beans
− Costs	—	$ 5.00 for fertilizer, seeds, etc.
Net (what the farmer actually keeps) income	—	$ 5.00 profit for the farmer

(Of course, in actual farming, we are talking about thousands of dollars.)

One of the largest costs in modern farming is machinery. Land was once the most important investment. But as farmers produce more crops on less land, what is most important is how each farmer uses the land, not how much land he or she has.

When farmers prepare their budgets, they look at the supplies they will use, how much they expect to produce, the costs, and any possible problems they can predict. Three basic records guide the farmer in running the business. These records are essential to the modern farmer. They may be a little difficult for the

A *financial statement* shows:

- how much money the land is worth
- how much the farmer owes for loans and bills
- what other things the farmer owns that are worth money, such as machinery
- how much ready cash is available

A *profit and loss statement* shows:

- the farmer's sources of income, such as the product
- how much income is earned
- what it costs to operate the farm

A *cash-flow statement* shows:

- the farmer's costs and income on a monthly or daily basis. This statement guides the farmer each day, indicating whether costs and income are reasonably well balanced.

nonfarmer to understand, but they help to give us some idea of how complicated farm life has become.

For instance, a farmer who wanted to produce livestock would have to consider as costs: buying the animal, corn, silage, protein to add to feed, hiring labor, transportation, marketing, power (energy), equipment, veterinary costs, and how much less a breeding animal might be worth in the future.

Most often *income* is determined by a farmer's skill, the market price of his produce, the quality of the land, the quality of the livestock, and the cooperation the farmer gets from the weather. No farmer can control all these elements, so farming is a blend of science and luck.

Compared to other people in business, the farmer does not earn a tremendous amount of money. In 1980, this was what an average farmer's earnings looked like:

$ 2,883	on-farm income
$ 4,826	nonfarm income
$ 7,709	total

Many farmers not only work on their own farms, but have to work at additional jobs to make a reasonable living. Farmers add to their income by renting parts of their property and by making various investments. They also receive financial assistance from the federal government.

MARKETING— GETTING THE FOOD FROM FARMER TO CONSUMER

In the 1800s there was no need for a marketing system. The farmer grew his wheat, took it to the mill, and brought it back to use as flour on his farm. Over the years farmers have produced more and more. They began to have a *surplus*—more than they could sell. They had to trade their surplus with others to have what they needed.

Out of this basic kind of trading grew the need for *middlemen* to collect, transport, store, process, and sell these products for farmers. (Today there is an entire industry called "agribusiness," which includes these people who are involved with farm products "off the farm": equipment makers, food processors, transporters, and wholesale and retail marketers.) Marketing involves many activities, starting with goods when they are first made, and ending when they are consumed. We will discuss this process in two parts: the way marketing works, or its functions; and the markets themselves.

The way marketing works is twofold: first, farm products must be processed. Cotton must be made into fabric; wheat must become a loaf of bread. Second, farm products have to be distributed in a steady supply to consumers. (For example, we want to be able to buy apples all year, not just at harvest time.)

There is no overall, organized marketing system. Rather, there is a variety of different, connected systems. Marketing has three purposes: to exchange goods, to handle goods, and to serve the needs of the farmer and the public.

- *Exchanging* means buying and selling. For the farmer this means getting the highest possible price. For the buyer (the middleman) this means selecting the best product.

- *Handling* means assembling the product; grading or standardizing (as with eggs or meat); storing in grain elevators or warehouses; transporting by train, truck, or air; and dividing and packaging the product. Part of handling is *processing*, or changing the form of the product, such as changing wheat to flour.

- *Services* such as financing or providing information about the market are important to farmers and businessmen alike. Middlemen borrow money from banks to pay farmers. Information about the current price of items being traded, or about weather conditions, is vital. Most market information comes from the U.S. Department of Agriculture.

The *markets* themselves are where a product actually changes ownership. When people buy or sell, the forces that determine the price paid are called *supply* (how much the seller has) and *demand* (how much the buyer wants). If the year has produced a good corn harvest, the prices will probably be low, since there is a bigger supply but the demand may be the same. If there was not much corn produced, farmers can get a higher price for what they sell.

The biggest markets are the *wholesale* markets. Wholesalers buy from farmers, store goods, and then sell them. Wholesale markets usually are near major transportation such as trains or ships.

Wholesalers sell their goods to *retailers*, who then sell to the public. Supermarkets and restaurants are retail markets for food products.

In *direct markets*, such as roadside stands and farm markets, farmers either sell their own produce to the consumer or sell it directly to the owner of the market or stand. With this arrangement, consumers enjoy the benefit of fresh food, usually at a considerably lower price.

INVESTMENTS

Farmers and other business people invest their money. Because the price of a product may change, the farmer often uses *futures contracts*, in which he is guaranteed a particular price.

Frozen foods on a grocer's shelf represent the product of many people's work—farmers, food processors and packagers, transporters, wholesalers, and finally the retail grocer.

Using a futures contract, a farmer and a buyer (for example, a canning firm) agree that the farmer will deliver several pounds of peaches on a certain date. The canner agrees to pay the farmer a set price for the fruit. This practice, which involves a commitment on both sides, reduces the risk a farmer would encounter trying to sell the fruit "on the spot." He is not so dependent on current prices.

Some products that are traded using futures contracts are corn, wheat, cotton, sugar, and oats. They are called *commodities*, which means they are items that can be shipped for trading purposes. Today almost any farm product falls in this category.

Based on trying to predict what commodity prices will be in a given year, business people become members of a *commodity exchange* to try to make profits based on their predictions. Exchanges provide a place for these people to buy and sell. Commodity exchanges are considered risky, because farmers can have good or bad years depending on weather, insects, or disease. The major exchanges are in Chicago, Kansas City, Minneapolis, New York, New Orleans, Los Angeles, Milwaukee, and Memphis.

EDUCATION AND CAREERS

One of the most important sources of information for the farmer is the *county extension service*. Extension services were established by the government in 1914 with the idea that county agents would travel around and give farmers practical advice. These services are operated in connection with the *land-grant colleges* that are our main base of agricultural education. In addition, there are many vocational-agricultural courses offered at the high school level. Each land-grant college runs an *experiment station* for research purposes.

Jobs are available in fields such as agronomy (crop science), agricultural economy, nutrition, soil science, and veterinary medicine. The government employs people to do research, to inspect crops and livestock, and to analyze farm economics. Businesses and schools hire teachers, and specialists in marketing, farm real estate, and farm loans. Some experience is gained right on the farm, but many successful owners, operators, and managers of farms also go to agricultural colleges to learn new skills.

THE FARM ECONOMY

People are today very concerned about the state of the farm economy. In the 1970s, there was a large decline in the income of farmers, due to several problems, some of which were related to general problems in the U.S. economy.

One problem was that new farmers were investing heavily in land and equipment but were unable to repay their loans because they didn't make the amount of money they had expected. In 1974, crop prices rose, leading to high feed prices and lower profits for livestock farmers.

Serious damage from drought reduced the farmers' harvests in 1977. So the late 1970s and early 1980s have been a time during which farmers have tried to bring about their own economic recovery. As part of this effort, farmers have increased their dependence on the government for financial assistance, as we see in the following chapter. Farmers must use this aid wisely and also try to lower their debts in order to improve their economic situation.

7

ORGANIZED FARMING

Farming has become more of an organized enterprise than it once was—some of this change has been due to the government, and some of it to the farmers' own efforts to organize. We will look at four broad areas separately, although they overlap with one another: The U.S. Department of Agriculture, government programs, various laws that relate to farming, and farmers' organizations.

THE U.S. DEPARTMENT OF AGRICULTURE

The Department of Agriculture (U.S.D.A.) was created by Congress in 1862 under President Lincoln. This once-small department collected agricultural information, distributed seeds, and published pamphlets.

Modern farmers depend on the U.S.D.A. for many things: for

basic needs, such as animal and plant research, as well as for conservation techniques, and food inspection. But the U.S.D.A. offers a tremendous variety of services, for farmers, students, and others wishing to learn about farming.

We will briefly discuss these services so the reader may get some idea of the enormous scope of the U.S.D.A. It is particularly important to realize that the government has a great impact on a farmer's life. The government serves the farmer, and also places limits or standards on farm products.

Some of the greatest contributions of the U.S.D.A. have been in the area of research, including pest and disease control, conservation, and education. Other important functions of the U.S.D.A. are economic assistance to farmers and the establishment of standards that ensure fair practices in the marketplace. The U.S.D.A. also deals with current problems such as the development of alternative energy sources, providing adequate food for people in America and in other countries, and helping farmers to improve the quality of life in their homes and communities.

GOVERNMENT PROGRAMS

In addition to, and in combination with, the various offices of the U.S.D.A., the government has many programs to help farmers. These programs were started because the government recognized that our nation's economy depends greatly on farmers and their ability to produce.

Today the amount of financial support farmers receive from the government is at a remarkable high of over $20 billion. Although citizens and politicians recognize that farmers need help, they are concerned about this high level of government spending.

Price supports started in the 1930s during the presidency of Franklin Roosevelt. Farmers were suffering because their incomes were so low. The government began to support the farmer against the low price he was getting in the marketplace by setting a "fair price" for his goods.

Today price supports take many forms. The goal of these supports is to make it possible for the farmer to continue to produce, and to keep the overall farm market in a stable position. Programs are managed through the U.S.D.A.'s Agricultural Stabilization and Conservation Service, the Commodity Credit Corporation, the Farmers Home Administration, and the Federal Crop Insurance Corporation.

Farmers receive support through a variety of ways, among them *loans, subsidies, reserves, diversion* money, and *payment-in-kind.* Farmers can get *loans* from the government at a very low rate of interest. Loans are given for day-to-day farm operations and for emergency use.

Subsidies are payments made to farmers when the market price for a particular product is very low. They ensure a reasonable profit for the farmer.

The government has also set up a program of *reserves* especially for grain. As farmers have produced more and more grain, they have ended up with too much—a surplus they can sell only at a loss. The government steps in and buys the grain rather than allowing prices to fall. The federal government stores the grain and uses it in various ways: to be released to the market when the

A soil conservation expert from the U.S.D.A (left) studies a map with a Texas rancher who is participating in the Department-sponsored program.

supply gets low or to send as food for needy people at home and in other countries.

Diversion programs are also used when farmers are producing a surplus. The government pays farmers to use their land for other purposes than growing crops. The land can be used for grazing or for such conservation measures as planting trees.

The problem of farm surpluses has most recently been dealt with by a program begun in 1983 under President Ronald Reagan called the *Payment-in-Kind (PIK)* program. This program has involved wheat, corn, cotton, and rice.

In the fall of 1982, programs were already in effect that diverted land for uses other than cultivation. Then farmers had an unusually good year of grain production, and there was not enough grain being purchased both in the U.S. and overseas: it was a case of too much supply and not enough demand. So PIK was introduced.

PIK means that farmers are "paid" with grain instead of cash. The PIK program tries to provide a better balance between supply and demand, increasing the farmer's income while decreasing the level of cash coming from the government.

LAWS

Since the early years of our nation, Congress has enacted agricultural legislation. Many of the programs just mentioned were created by law. It is easiest to view these major laws by looking at some of the highlights of the history of farming legislation:

1796 ▪ *Land Act* passed to provide for sale of public land. Revised many times. ▪ *National Board of Agriculture* set up by George Washington.

1862 ▪ *Department of Agriculture* established by Abraham Lincoln. ▪ *Homestead Act* opened up western land by offering it at low cost, or no cost if farmed for five years. ▪ *Morrill Act* provides for *land-grant* colleges.

1877 ▪ *Desert Land Act* allowed settlers to buy certain land if they would irrigate it.

1887 ▪ *Hatch Act* officially established experiment stations to promote research.

1890 ▪ *Second Morrill Act* created land-grant colleges for black students.

1907 ▪ *Meat Inspection Act* created inspections for animals and meat for import and export.

1933 ▪ *Agricultural Adjustment Acts* passed to help farmers economically. ▪ Commodity Credit Corporation created to provide storage for surplus goods. ▪ Farm prices reset to better rates.

1935 ▪ *Rural Electrification Act* enabled farmers to get loans to electrify their farms.

1937 ▪ *Agricultural Marketing Agreement Act* passed to maintain fair marketing practices.

1946 ▪ *National School Lunch Act* provided food aid to school children.

1964 ▪ *Food Stamp* program aided poor families.

1965 ▪ *Food for Peace Act* renewed U.S. pledge to aid other countries in need of food.

Much surplus food produced by American farmers is bought by the government for school lunch programs.

1973 ▪ *Agricultural and Consumer Protection Act* focused concern on conservation.

1981 ▪ *Farm Act* passed to grant cash supports for wheat, corn, rice, and cotton.

FARM ORGANIZATIONS

The other part of organized farming, apart from the government's involvement, has to do with the farmers themselves. The 1800s and early 1900s saw the development of farm *cooperatives*, associations the farmers set up to help market their goods. As an organized group, farmers were able to purchase supplies at better rates, get better prices for their products, and make agreements with middlemen.

Cooperatives, also called "co-ops," gave the farmer more power in the marketplace. In some cases, such as with milk and oranges, co-ops have also set up their own processing or distributing areas.

Other groups called *commodity organizations* became very important in the 1960s. These organizations represent farmers of certain commodities such as grain or milk, and try to promote laws that favor them.

In general, farmers are not very political, except that they tend to vote for officials who are in favor of good farm policies. But they can become political when necessary. Farm organizations represent many people and carry a fair amount of political weight, especially during election years. Presidential candidates make a great effort to get the "farm vote."

Certain organizations represent the general interests of the farmer. Some of these organizations work in Congress to get laws passed that are favorable to farmers. They are the American Farm Bureau Federation, the National Farmers' Organization, the National Farmers' Union, and the National Grange.

Other organized farm groups are labor unions, which represent the hired workers. The most famous of these is the United Farm Workers. A Mexican-American labor leader, Cesar Chavez, organized the union to improve the situation of migrant workers, hired workers who move from one farm to another. Through strikes and boycotts, this union has become nationally known.

8

THINGS TO THINK ABOUT

All through this book there have been many facts; but there have also been some questions raised about the nature of modern farming. In this chapter we will emphasize some issues in the hope of understanding some of the complex questions that farmers, consumers, and politicians face. Some of these will be mentioned only briefly, since they have been addressed elsewhere in this book.

Automation and technology ▪ One of the greatest events in farming has been the ability to mass-produce food items. This achievement has earned America the nickname ''breadbasket of the world.'' First the tractor, then electricity, gave farmers the power to produce an enormous amount of food on a small amount of land. However, technological advances raise some important questions.

First, automation has meant that machines do the work that

people once did—so there are fewer people needed to work on farms. While this makes farming a more efficient operation, it takes some of the "home-style" flavor out of farming. Also, farmers must be experts with extensive training and education.

Second, technology cannot entirely defeat the unpredictable elements of farming such as the weather. Drought, floods, and even new strains of disease can still destroy the farmer's work.

Third, machinery requires a large financial investment. Some solution must be found so that struggling beginners do not have to become deeply in debt in order to become farmers.

Sophisticated scientific techniques—whether they include machines, chemicals, or new methods of breeding—must be coordinated with the basic elements of farming. Farmers must use them carefully so as not to upset the balance of nature.

Recently, the U.S.D.A. has made strides that address farmers' concerns. Work is being carried out with computers to help figure out how to control land and water erosion. Also, attempts are being made to improve monitoring of the weather, which will help farmers greatly.

The U.S.D.A. is doing additional major work to prevent pest and disease destruction of crops and livestock. With breeding experiments, improved lines of plants and animals are stronger. And new chemicals have been developed to destroy insects.

Conservation and pollution ▪ As we have seen in our discussion of erosion, continued efforts must be made to use land and water carefully, and to prevent their abuse and pollution. Pollution endangers our food sources. We need to investigate carefully the chemicals we use in farming. In 1972, the USDA banned DES (diethylstilbestrol), a cattle-feed chemical used to stimulate growth. Some meat from cattle that had been given this chemical was found to have a possible danger of cancer. The ban was later lifted for certain cases, but this case points up the need for careful use of chemicals.

Energy ▪ U.S. farmers use over one-fifth the total energy of our whole economy. Therefore, there is much work being done to find new, less expensive means of power for farming.

Major work is being done in these areas:

Solar energy will probably be limited to regions where there is a dependable amount of sunlight. Potential uses are for grain drying, heating of animal shelters, heating and cooling of rural homes, and pumping water.

Biomass is organic matter produced by plants and animals. It is an inexpensive source of energy because it is produced on farms. Research is being done to convert biomass to alcohol, for use as fuel.

Windmills can be used to pump water and produce electricity. Because the wind is unpredictable, wind power will be most useful in combination with other systems. It can be used for powering irrigation systems and heating or cooling buildings.

Geothermal energy comes from the hot water located deep within the earth which is pumped upward. It is useful for greenhouses, animal care, drying crops, and fish farming.

Hydroelectric power uses water to power electric motors. The United States has several dams already in operation. They can be expensive but are useful if designed efficiently.

All of these new sources of energy are being cautiously examined.

Probably they will be used along with, not in place of, the petroleum supplies that are our present major energy source. But at least they represent positive alternatives for the future.

Labor relations ▪ Partly due to the increased use of machinery but also to the need to hire migrant workers, farmers are facing some new difficulties. The problem of getting farm work done and still maintaining good, fair relationships with workers has yet to be solved. Especially in the southern sections of states which border on Mexico, this issue is difficult, since many workers come into this country, sometimes illegally. Workers are being treated somewhat more fairly than in the past.

The world food crisis ▪ Since the early 1950s the need for food throughout the world has more than doubled. Even with the tremendous increase in food production, the world population growth is so rapid that, especially in the developing countries of Africa, Asia, and Latin America, starvation is a constant and very real threat. In this book, we include a section dealing with this troublesome problem for two reasons. First, because America produces an abundance of food, it therefore bears some responsibility for solving this problem. Second, American farmers produce the surplus that can be used to help feed other countries.

In addition to the problem of world overpopulation, we in the United States (as well as other wealthy countries) actually con-

Surplus grain from the United States is distributed here in an emergency relief program in the East African nation of Uganda.

sume more food than others. We are used to a wide variety of foods. Also, due to the control of disease, people are living longer, and the survival rate for newborn babies has increased, so there are more people in the world to feed.

The United States started a program in 1954 called Food for Peace, through which food was donated at low or no cost to needy countries. Food has also been distributed through the Food and Agriculture Organization of the United Nations. The World Food Conference, held in Rome in 1974, was set up specifically to deal with this crisis.

The conference dealt with emergency feeding, but also some more sophisticated issues: making the food systems of the wealthier nations less wasteful, making energy sources available to poorer countries, and distributing food to all nations in need, not just those who were politically "friendly." The conference accomplished the following results: it set up an international grain reserve and a detection system to warn countries of shortages of food; it established a World Food Council under the United Nations; and it urged the United States to contribute more food, which it agreed to do.

In addition, other solutions have been attempted, such as new ways to grow food. So-called "miracle" strains of wheat and rice were developed in the late 1960s and early 1970s, and although these new varieties were less than completely successful, better crop yields were achieved. Also, Americans are switching from beef to other, less expensive, protein sources, such as chicken.

There have been many attempts to teach other countries to farm in a more modern way, so that they can begin to move toward self-sufficiency in food. This is done through the United Nations as well as the United States Peace Corps.

Other important concerns within other countries will be an increased and responsible awareness of the problems of over-population. In some countries, where farms have been owned either by the government or by a small group of powerful land-owners, farmers are now being allowed to own land, which gives them an incentive to produce more food.

The problem is by no means solved. The years to come will see a continuation of this food crisis, and it will be a very pressing problem.

Farmers will, of course, continue to work toward resolving all of their important concerns. But as responsible consumers and citizens, those of us who are not farmers can also help with some of these issues. First, we can be aware of the importance farming has for us, and educate others as well. Second, we can support organizations and legislation that protect our natural resources for agricultural use. And last, we can share our farming knowledge and techniques with other nations to help them gain greater levels of self-sufficiency.

GLOSSARY

breeding: producing animals or plants. Selective breeding is done to produce special characteristics.

cash crop: a crop that is raised for sale.

climate: the general weather conditions of an area.

combine: a machine that combines the work of the harvester and the thresher.

commodities: items that are transported for sale.

commodity exchange: a place where farmers and businessmen invest money and try to predict crop production.

commodity organization: groups that band together to try to get a good price for their product.

conservation: efforts to save natural resources, especially land and water.

cooperatives: farmers' groups that buy supplies and sell their products as a group instead of as individuals.

costs: expenses that a farmer must consider.

cotton gin: a machine that separates cotton fiber from seed.

county extension service: government office responsible for helping farmers through information and technical services.

crop rotation: planting certain crops in sequence with others to put nutrients back into the soil.

direct markets: markets such as roadside stands, where the public buys produce from the farmer or from the owner of the stand.

diversified (mixed) farms: farms that produce several crops or kinds of livestock.

diversion programs: government-supported efforts to conserve land by using it in other ways than for producing crops.

dry farming: a method that uses rain for needed moisture but does not use irrigation.

erosion: a process of wearing away, especially of the soil that lies on the surface of the ground.

experiment stations: places set up in connection with land-grant colleges for research.

fallow: cultivated land that is not planted.

feed concentrates: chemicals added to animal feed to help livestock grow or produce better.

fertilizer: material added to the ground to help crops grow.

Food for Peace: program that helps feed hungry people in other countries.

forage: livestock food such as grass.

futures contracts: agreements which farmers and middlemen make for purchase and future delivery of goods.

futures trading: investments based on futures contracts.

grain reserve: grain held by the government to keep prices stable and, in the event of a poor grain harvest, to prevent a shortage.

harrows: machines used to till the soil.

herbicides: chemicals used to kill weeds.

hybrid: a plant that has been produced by selective breeding; a mixture of certain qualities of other plants.

income: the amount of money a farmer makes.

insecticides: chemicals that kill insects.

irrigation: a process that helps farmers to water crops.

land-grant colleges: colleges started to further the progress of agricultural knowledge.

middlemen: business people who buy goods from the farmer and sell them to others.

migrant workers: hired workers who move from one farm to another depending on when the harvest is due.

Payment-In-Kind: also called PIK, a program to pay farmers with grain instead of cash; used by the government to avoid a grain surplus.

plow: a tool used to dig up the ground.

pollution: a situation where the environment (air, water, or earth) has become unhealthy for animal or human life, usually due to chemical wastes.

price supports: fair prices paid to the farmer by the government to improve the farmer's economy.

processing: various stages that occur after a farm product has been sold; includes freezing, canning, packaging, and other phases that change the form of the product.

reaper: a machine that cuts grain and, with certain attachments, can also stack or collect it.

retail markets: markets that sell to consumers, such as supermarkets or restaurants.

seed drill: a tool that drills holes in the soil for seeds.

silage: feed that has been stored and moistened in a silo.

soil structure: the texture of the soil; its ability to hold water and support plant growth.

specialized farms: farms where one or two major crops or kinds of livestock are produced.

strip cropping: the practice of planting crops in strips to enrich the soil and prevent erosion. Usually crops are planted between grasses.

subsidies: loans and financial support given to farmers by the government.

subsistence farming: farming for survival rather than profit.

supply and demand: a business theory that says the amount of what is available in the marketplace determines its price.

surplus: overproduction of a certain good.

thresher: a cutting machine used to harvest grain.

tillage: the process of digging up the soil for planting; cultivated land.

topography: the shape of the land, mountainous or level; helps to determine what kind of crop or livestock a farmer will choose.

tractors: farm machines that changed farming greatly; used alone or to pull other machines.

wholesale markets: markets in which farmers sell to middlemen who in turn will sell to retailers; usually centrally located between farms and the retail markets.

World Food Conference: a meeting held in 1974 to restore an international commitment to feed people unable to feed themselves.

FOR FURTHER READING

Edwards, D. J. "Growing Food," in *Finding Out About Science.* Stella Robinson (ed.). New York: John Day Company and Rupert Hart, Davis Educational Pub. 1969.

Floethe, Louise and Richard Floethe. *Farming Around the World.* New York: Charles Scribner's Sons, 1970.

Helfman, Elizabeth S. *Land, People and History.* New York: David McKay, 1962.

———*Our Fragile Earth.* New York: Lothrop, Lee and Shepard, 1972.

McMillen, Wheeler. *The American Farm Story.* Canada: Holt, Rinehart and Winston, 1961.

Perl, Lila. *The Global Food Shortage.* New York: William Morrow, 1976.

Sullivan, George. *How Do They Grow It?* Philadelphia: Westminster Press, 1968.

World Book Encyclopedia, 1978. Entries on *Farms and Farming,* and on *Agriculture.*

PLACES TO VISIT

National Agricultural Library Building, Beltsville, MD. There is a twenty-minute self-guided tour with a tape-recorded message.

Farmers' Museum, Cooperstown, NY. A comprehensive display of farmers' tools and colonial farming.

Old Sturbridge Village, Sturbridge, MA. A working model of an early farming village with displays.

Smithsonian Institute, Washington, DC. A museum that features information about various areas, among them early farming.

Farm and ranch vacations for families or young people are described in the guide, *Country Vacations, U.S.A.,* c. 1976, pub. by Farm and Ranch Vacations, Inc., New York, NY.

INDEX

ABOUT THE AUTHOR

Jill Menkes Kushner is a freelance nonfiction writer whose articles have appeared in newspapers in New Jersey and Connecticut. A graduate of Boston University, Ms. Kushner is currently studying for a masters degree in corporate communications at Farleigh Dickinson University. In addition to her work as a writer, she has taught school and worked with emotionally disturbed children. Ms. Kushner lives in South Orange, New Jersey, with her husband, Robb, and her two children, Matthew and Elizabeth.